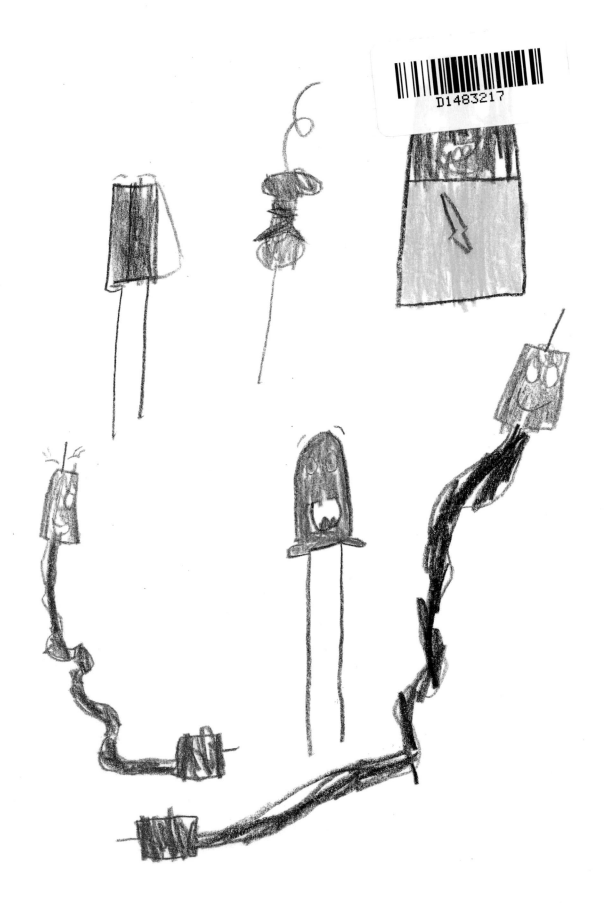

## CARROT PANTS

presents

# ED GETS HIS POWER BACK

Written by
**Brandon Satrom**

Illustrated by
**Jake Page**

# For Sarah, Benjamin, Jack & Matthew

First Edition 2017

Library of Congress Catalog Card Number pending

ISBN 978-0-9991576-0-2

Printed in China

This book was typeset in Futura.

The illustrations were created digitally.

Endpage illustrations were created with Lyra Rembrandt Polycolor pencils.

Carrot Pants Press

4500 E. Palm Valley Blvd Ste. 108-108

Round Rock, TX 78665

visit us at www.carrotpantspress.com

This is Ed.

Ed is a Light-Emitting Diode, or LED for short.

That's a fancy term for something that lights up
when you give it power.

Ed lives in Breadboardeaux with all his friends.

Ed the LED has a round head and a big smile. He's a happy fellow, and doesn't even mind that one of his legs is longer than the other.

LEDs have a special job in Breadboardeaux. They light up the city at night.

And when Ed feels the city's power flowing through him, he shines a bright, green light for all to see.

Ed's family is responsible for lighting up the neighborhood. Today is Ed's first day doing this special job while his parents are away for the evening.

And today, this job is more important than ever. You see, the power is out in Breadboardeaux and the sun will be setting soon.

# Ed the LED has lost his power.

"Oh no," Ed said. "I have lost my light! The power is out in the city and the sun is setting soon! My friends need me! What can I do?"

"I know! Patty can help me!"

Ed bounced down the street until he reached Patty's house at Avenue B and 7th street.

"Hi Patty," Ed said.

"Hello Ed," Patty said. "Where is
your green light?"

"I've lost my light," Ed said. "The power
is out, and the sun is setting soon!
Can you help me?"

Patty the capacitor smiled.

You see, Patty is also special in Breadboardeaux.

She can hold onto energy like a battery
until her big cylinder-shaped body is full. Then,
she can send that power to her next friend
down the line.

"I can help you, Ed," Patty said. "Get in line
next to me and let's get your power back!

Patty had plugged her feet into the street
at Avenue B. One leg sat at 7th Street and
the other at 9th Street. The streets of
Breadboardeaux carry power
under the ground. If two friends
plug into the same street,
they can pass power
to each other!

"Oh, thanks Patty!" Ed said.

Ed jumped to Avenue E, a few blocks over from Patty. He plugged his legs in at 7th Street and 9th Street, first his long front leg, then the back, short leg. Patty's power should flow out from 7th Street right over to Ed.

"Ready!" Ed said.

Patty closed her eyes, twisted up her face and concentrated on sending her power to Ed.

But there was no light. Patty sighed.

"I'm sorry Ed," she said. "It looks like I'm all
out of power."

"Oh no!" Ed said. "But the power is out in the city
and the sun is setting soon. What can we do?"

"To get you some power, I need power," Patty said. "And to get power for me, we need a power source."

"A power source?" Ed asked.

"Yes. I am not a source of power, I store it up for short periods of time, and then I'm all out. We need a battery."

"A battery," Ed said, as he scratched his head.

Patty smiled and pointed down the street.

"Here comes Zara! She can help us."

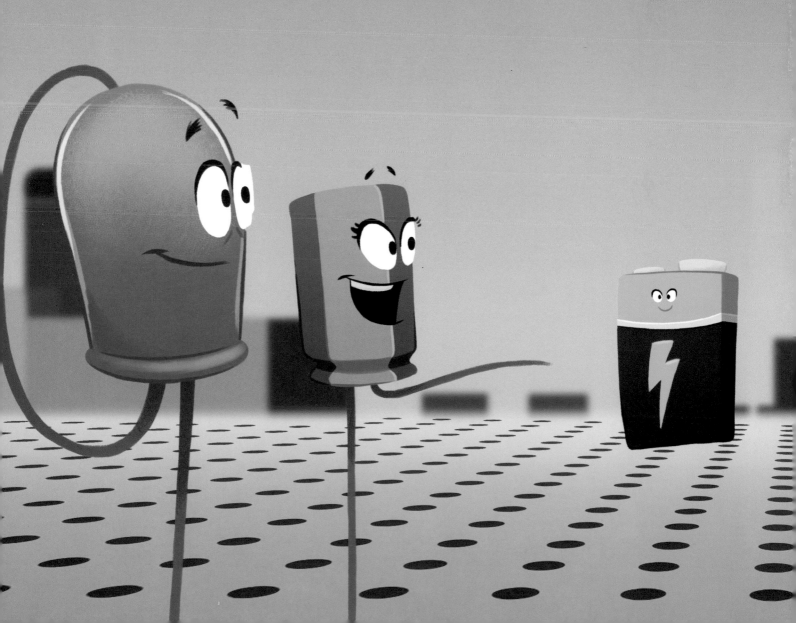

Zara Bolt is a 9-volt battery who packs a lot of power. Plenty to get Ed's light shining again.

"Hi Ed. Hi Patty," Zara said. "What seems
to be the trouble?"

"I've lost my light," Ed said. "The power is out and the sun
is setting soon. I came by to see if Patty could help."

"But I don't have any power left," Patty
said. "Can you help?"

Zara smiled a big smile and spun around
like a quarter on a table.

"You bet I can!" she said. "I'm fresh out of the
package and full of power!"

Zara jumped next to Patty and looked at
Avenue A and 7th Street one block over. There,
she saw a hole where she could send power to
Patty underground.

She looked at Patty and saw her feet in the holes.

She looked at Ed and saw his feet in the holes.

She looked down at herself and remembered she had
no feet for the holes.

"Patty, I have the power you need to give to Ed," Zara said. "But I have no feet and I need them to give you that power."

"Oh no," Patty said. "The power is out and the sun is setting soon! What can we do?"

Ed had an idea.

"Zara, doesn't your mom have a power suit she uses sometimes?"

"Of course!" Zara said. "The power suit! Be right back!"

Zara zipped away and reappeared wearing her power suit, a case that would help her share her power.

"Let's do this!" Zara said. "Now, where do I plug in?"

Zara looked to her left and right, and up and down. Her suit had places on the top where power could flow out of her. But she still had no feet, and no way to plug into the street.

"My suit has no feet!" Zara said. "The power is out! And the sun is setting soon! What can we do?"

Zara looked at Patty. Patty looked at Ed. Ed looked at Zara. All at once, they said. . .

"We know! The Gonzales twins!"

Zara zipped away again and returned with
Rita and Bobby Gonzales.

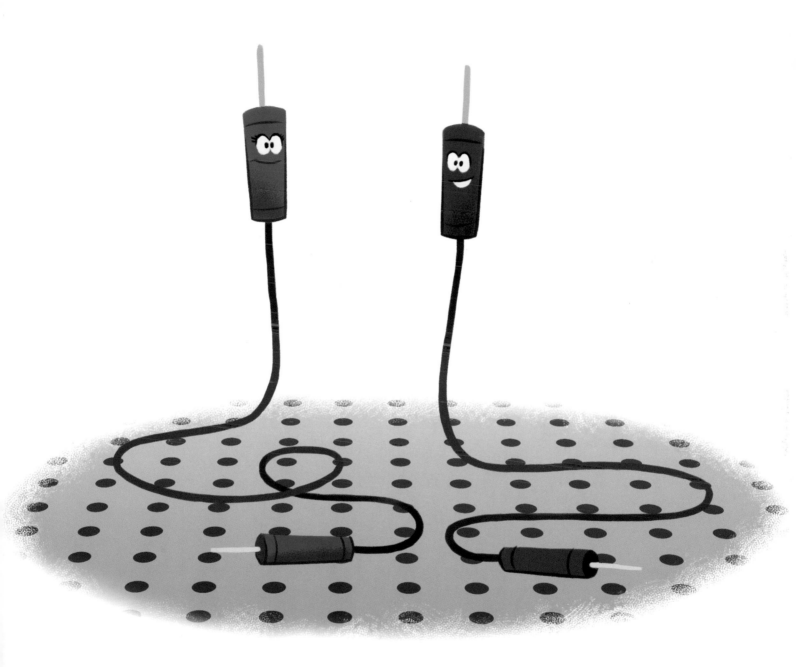

The Gonzales twins are red and black wires with little
feet at each end. With one foot, they can connect to someone
else, and plug their other foot into the street.

"We have everyone we need now!" Zara said.

"Plugging in now!" Rita and Bobby Gonzales said. Rita plugged into 7th Street, a block over from Patty, and Bobby plugged into 9th.

"Here I go!. . ." Zara said.

Everyone froze.

"Mr. Resistor!" Ed exclaimed.

"Hi kids!" Mr. Resistor said. "Sorry for shouting, but you were about to lose your light for good, Ed!"

"Oh no!" everyone said.

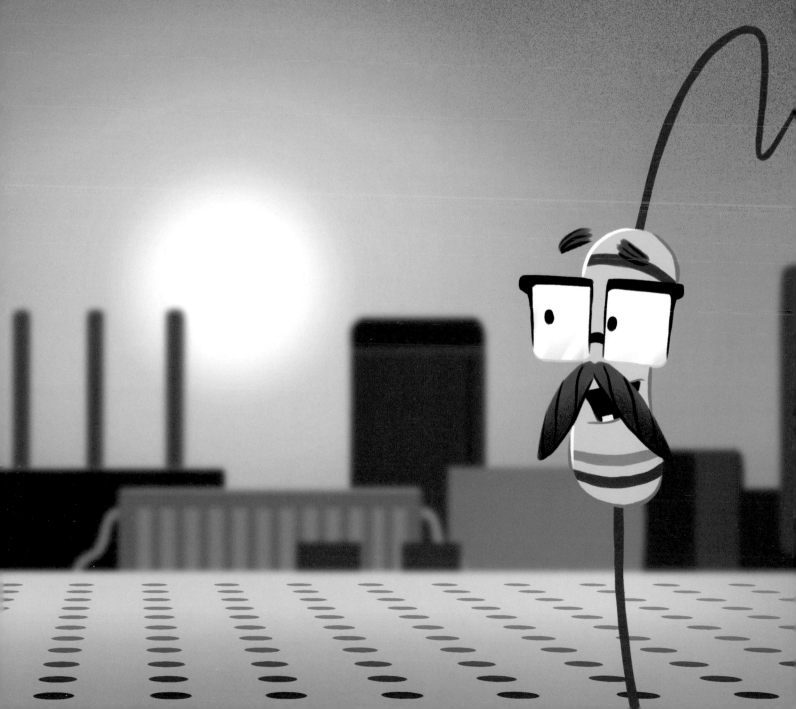

"Yessiree," Mr. Resistor said. "You see, Zara has plenty of power to give you. Too much, in fact. Patty can store it up and pass it along, but if you get too much power, your light will pop and never light up again."

"Thanks for saving my light, Mr. Resistor," Ed said. "But the power is out and the sun is setting soon. Can I ever get my light back?"

"Wow!" said Ed.

"Scoot on over and I'll show you."

Ed pulled out his longer leg and moved it to 8th Street,
while keeping his shorter leg in place at 9th.

"Perfect," Mr. Resistor said as he jumped in line between Patty and Ed. He plugged in one of his feet at Avenue C and 7th Street, and the other at 9th next to Ed.

"OK! Is everyone ready?"

"Ready," Zara said.

"Ready," the Gonzales twins said.

"Ready," Patty said.

"SO ready," Ed said.

"OK, take it away, Zara!" Mr. Resistor said.

Carefully, Zara connected one side of her suit
to Bobby and the other side to Rita.

"Power coming your way!"

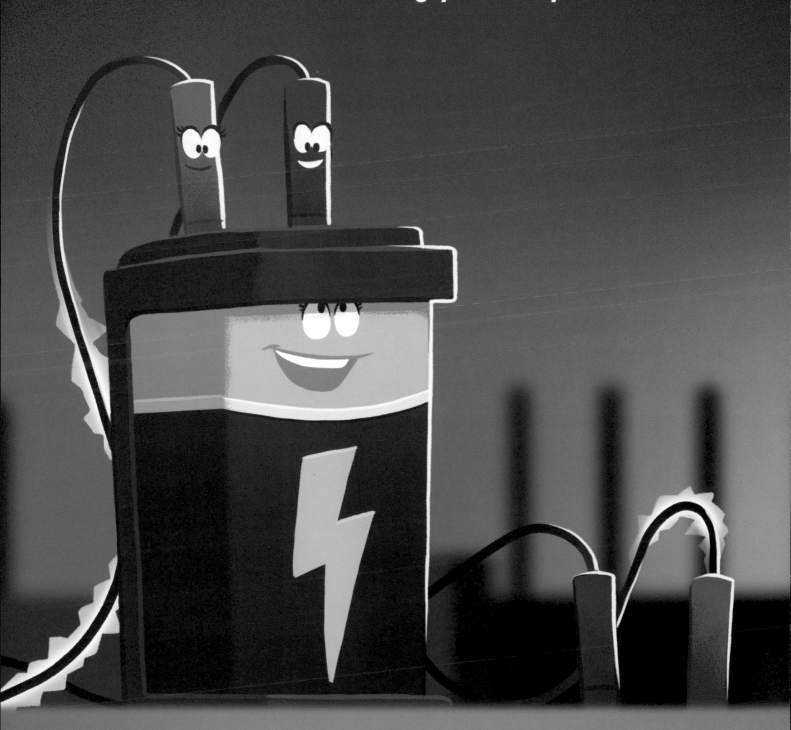

"I can feel it," Patty said.
"Here you go, Ed!"

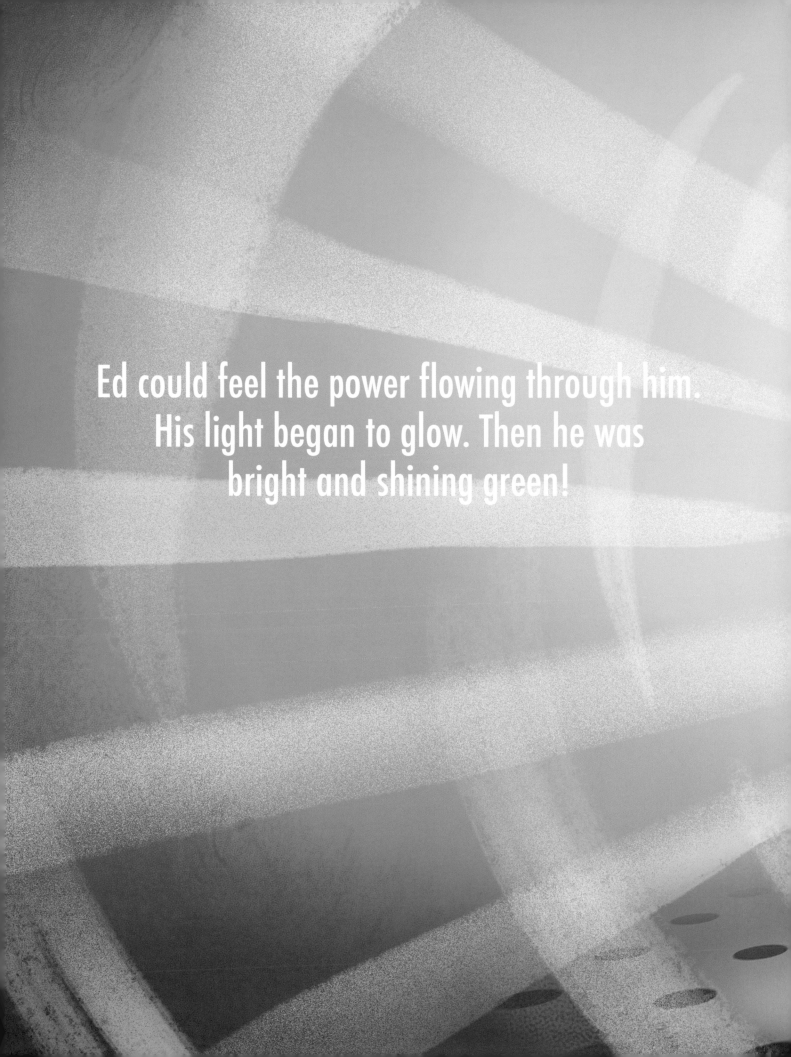

Ed could feel the power flowing through him.
His light began to glow. Then he was
bright and shining green!

"My power!" Ed said. "It's back!"

The city of Breadboardeaux was filled with a warm, green light. Everyone cheered as the sun disappeared from the sky!

# Ed the LED Will Return in...
## *Ellie Saves the Day!*

# Glossary

## Battery
A power source that provides electricity to electronic components. Batteries come in different shapes and sizes and power much of the world around us.

## Breadboard
A base used to build a circuit with electronic parts (components). Metal rails inside the base allow electricity to flow between components.

## Capacitor
A component that can store a small charge of electricity from a power source. This charge can be used to power other components even when a power source is disconnected.

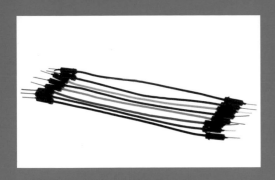

## Jumper Wire

A wire with a connector at each end that can be plugged into a breadboard to connect components in a circuit.

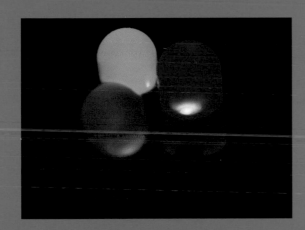

## Light-Emitting Diode (LED)

A component that shines a light when electricity passes through it.

## Resistor

A component that controls the amount of electricity passed along to another component. When used with LEDs, resistors reduce current to protect them from damage.

Thanks to all our Kickstarter Backers for making this book possible! Including...

Todd Anglin

Dwight & Wendy Ballard

John Bristowe

Erica Burns

Margaret & Don Carroll

Ryan & Rachael Chadwick

Sarah Dutkiewicz

Randy & Megan Evans

Chris Eyehorn

Cliff & Teri Heck

David Higgins

Burke & Erin Holland

Mehfuz Hossain

Betty Howard

Dimo Illiev

Albert & Sharon Liles

Chad & Johnna Liles

Shannon & Shanan Marek

Master Yang's World Class Tae Kwon Do

Matthew McNulty

Kelly Nagel

Justin Newcom

David O'Hara

Brian Prince

Clark & Carrie Sell

Courtney & Logan Stanley

Valio Stoychev

Tim G. Thomas & Cheyenne J. Clark

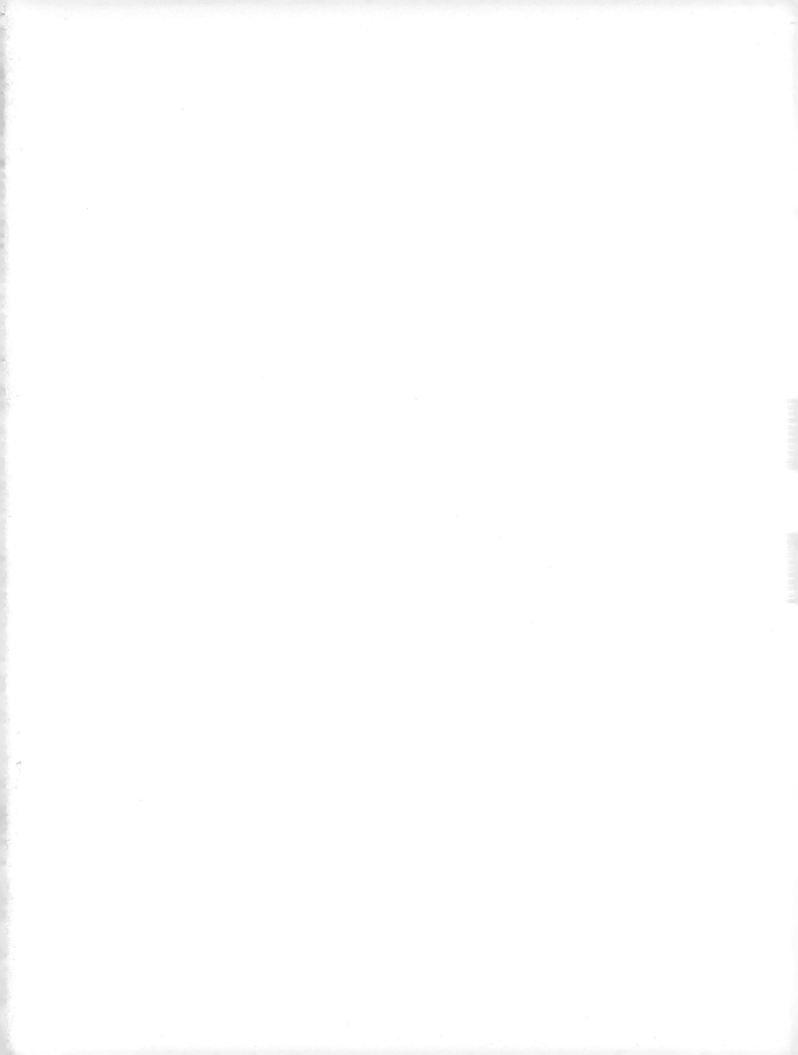

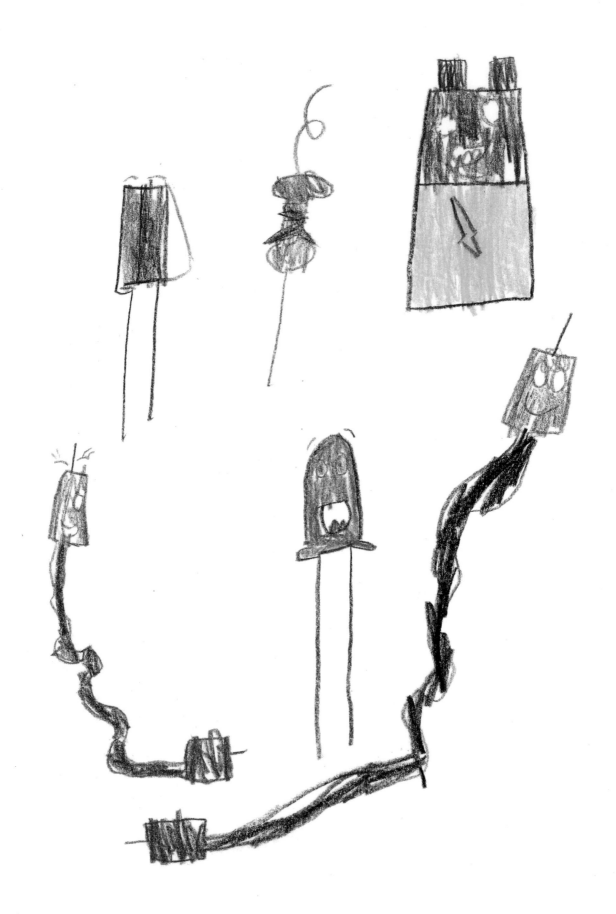